Grade 2

Spelling
PRACTICE BOOK

The ***McGraw·Hill*** *Companies*

Published by Macmillan/McGraw-Hill, of McGraw-Hill Education, a division of The McGraw-Hill Companies, Inc., Two Penn Plaza, New York, New York 10121.

Printed in the United States of America

11 021 09

Contents

Unit 1 • Relationships

Unit 2 • Growth and Change

Unit 3 • Better Together

Unit 4 • Land, Sea, Sky

Unit 5 • Discoveries

Unit 6 • Expressions

Name ______________________________________

Spelling Words

has	sat
wag	had
bad	fix
six	him
will	if

Using the Word Study Steps

1. LOOK at the word.
2. SAY the word aloud.
3. STUDY the letters in the word.
4. WRITE the word.
5. CHECK the word.
   Did you spell the word right?
   If not, go back to step 1.

Puzzle

Solve the puzzle. Circle the six hidden spelling words.

o	w	a	g	p	s	t
i	l	v	p	h	a	s
h	i	m	n	w	z	b
d	b	a	d	k	c	p
s	e	f	i	x	y	d
g	a	s	a	t	h	x

At Home: Review the Word Study Steps with your child as you both go over this week's spelling words.

Name ______________________

has	six	him	sat	bad
wag	if	will	had	fix

Word Sort

Look at the spelling words in the box. Write the spelling words that have the short *a* sound.

1. ________ **2.** ________ **3.** ________

4. ________ **5.** ________

Write the spelling words that have the short *i* sound.

6. ________ **7.** ________ **8.** ________

9. ________ **10.** ________

Misfit Letter

An extra letter has been added to each spelling word below. Draw a line through the letter that does not belong. Write the word correctly on the line.

11. hais ________ **12.** fixe ________

13. sayt ________ **14.** hyim ________

15. whill ________ **16.** iff ________

17. wage ________ **18.** baid ________

19. sixe ________ **20.** hayd ________

Name ____________________

has	six	him	sat	bad
wag	if	will	had	fix

Match-Ups

Draw a line from each spelling word to its meaning.

1. wag	to make better
2. six	the past tense of *sit*
3. fix	the number after five
4. bad	to move back and forth
5. sat	not good

Sentences to Complete

Write a spelling word on each line to complete the sentence.

6. I hope you ______________ come to see my class tomorrow.

7. Can I help ______________ with the test?

8. We ______________ a new teacher last week.

9. Now Ted ______________ a book on his desk.

10. What happens ______________ my bus is late?

Name ____________________

There are six spelling mistakes in the paragraph below. Circle the misspelled words. Write the words correctly on the lines below.

I haid a very good day at school. I got to help fikx the fish tank. Then Mr. Dan and I sayt down. I told hime all about a book I had read. He asked iff he could borrow it. I said, "Yes, I wil bring it in so the whole class can read it."

1. ____________ 2. ____________ 3. ____________

4. ____________ 5. ____________ 6. ____________

Writing

Write about your day at school. Use three spelling words from your list.

____

____

____

____

Name ______________________________

Using the Word Study Steps

1. LOOK at the word.
2. SAY the word aloud.
3. STUDY the letters in the word.
4. WRITE the word.
5. CHECK the word.
   Did you spell the word correctly?
   If not, go back to Step 1.

Spelling Words

went	not
tell	tug
pet	hut
job	tub
fog	bun

Find and Circle

Circle the 10 hidden spelling words. The words are across, down, and on a slant.

x	s	j	o	b	t	h	p	z	y	n	s
n	t	f	l	u	c	u	t	e	l	l	q
v	o	w	g	n	x	t	f	b	t	u	b
g	z	t	s	w	e	n	t	k	z	f	r

At Home: Review the Word Study Steps with your child as you both go over this week's spelling words.

Name ______________________________

went	tub	not	tug	fog
pet	tell	hut	job	bun

Word Sort

Look at the spelling words in the box. Match each word to a vowel sound. Write the words on the lines.

Short *e*	Short *o*	Short *u*
1. ________	4. ________	7. ________
2. ________	5. ________	8. ________
3. ________	6. ________	9. ________
		10. ________

New Words

Make a new word from the spelling list by changing the vowel.

11. pat – a + e = ________

12. bin – i + u = ________

13. jab – a + o = ________

14. fig – i + o = ________

15. hat – a + u = ________

Name ______________________________

went	tub	not	tug	fog
pet	tell	hut	job	bun

Write a spelling word to complete each sentence.

1. Tom lets me ______________ his dog.
2. I can ______________ that Jen likes me.
3. Gus and I ______________ to the park.
4. It is ______________ nice to be mean.
5. I bathe my puppy in the ______________.
6. Mike and Dan made a ______________ to play in.
7. I did a good ______________ helping my friend.
8. At dinner Dad gave me a hamburger on a ______________.
9. My dog likes to ______________ on his leash.
10. The ______________ makes it hard to see.

Name ____________________

There are six spelling mistakes in the letter below. Circle the misspelled words. Write the words correctly on the lines.

Dear Ben,

I want to tel you about Sam. He is my new peet rabbit. I weint to the store and got him a cage and some food. He is nat very big yet. I gave him his first bath in the tuab. Taking care of him is a big jaub. Please come see him soon!

Your friend,

Matt

1. ____________ 2. ____________ 3. ____________
4. ____________ 5. ____________ 6. ____________

Writing

Write a letter to a friend. Use three spelling words from the list.

____

____

____

____

Name ____________________

Using the Word Study Steps

1. LOOK at the word.
2. SAY the word aloud.
3. STUDY the letters in the word.
4. WRITE the word.
5. CHECK the word.
   Did you spell the word right?
   If not, go back to step 1.

Spelling Words

bag	mad
cap	back
ham	cape
bake	made
ate	rake

Sounds the Same

Write a spelling word that rhymes with the words in each group.

1. dad sad ____________
2. rate date ____________
3. lake make ____________
4. map lap ____________
5. nag rag ____________
6. ram jam ____________

At Home: Review the Word Study Steps with your child as you both go over this week's spelling words.

Name ______________________________

cape	bake	mad	bag	rake
ate	back	cap	ham	made

Word Sort

Look at the spelling words in the box. Write the spelling words that have the short *a* sound.

1. ____________ **2.** ____________ **3.** ____________

4. ____________ **5.** ____________

Write the spelling words that have the long *a* sound.

6. ____________ **7.** ____________ **8.** ____________

9. ____________ **10.** ____________

Word Find

Find and circle five spelling words in the puzzle.

a	b	q	p	t	k	e
a	m	a	c	x	u	d
t	a	i	g	f	y	l
e	d	c	q	r	s	h
y	e	g	b	a	c	k
u	q	x	v	k	x	j
z	t	l	w	e	m	n

Name ______________________________

cape	bake	mad	bag	rake
ate	back	cap	ham	made

Questions

Write a spelling word to answer each question.

1. What can you use to make a pile of leaves? ______________
2. What can taste good on a sandwich? ______________
3. What can you put on your head? ______________
4. What can you put your lunch in? ______________
5. What word means not happy? ______________

Sentences to Complete

Write a spelling word to complete each sentence.

6. I ______________ an apple for lunch.
7. Dad ______________ a fire at camp.
8. We drove ______________ home after the show.
9. Kate wore a black ______________ on her back.
10. I can ______________ the bread in the oven.

Name ____________________

There is one spelling mistake in each sentence. Circle the misspelled words. Write the correct words on the lines below.

1. Dad and I got a baig.
2. We went backe to camp with lots of sticks.
3. Dad made a fire, and we ayt fish for dinner.
4. I was maed when it started to rain.
5. We mayd sure the fire was out.

1. ____________ 2. ____________ 3. ____________

4. ____________ 5. ____________

Writing

Write about fire safety. Use three spelling words from your list.

____

____

____

____

____

Name ________________________________

Spelling Words

did	rip
fin	mix
pick	five
nine	side
pipe	hike

Using the Word Study Steps

1. LOOK at the word.
2. SAY the word aloud.
3. STUDY the letters in the word.
4. WRITE the word.
5. CHECK the word.
   Did you spell the word right?
   If not, go back to step 1.

X the Words

Put an X on the words with the long *i* sound.

rip	pipe	pick
hike	fin	mix
side	did	nine
five	lit	wick
fit	lick	sip
fix	bit	win

At Home: Review the Word Study Steps with your child as you both go over this week's spelling words.

Name ______________________

hike	did	rip	pipe	side
fin	mix	nine	five	pick

Word Sort

Look at the spelling words in the box. Write the spelling words that have the short *i* sound.

1. ________ **2.** ________ **3.** ________

4. ________ **5.** ________

Write the spelling words that have the long *i* sound.

6. ________ **7.** ________ **8.** ________

9. ________ **10.** ________

Sounds the Same

Write the spelling word that rhymes with each word below.

11. bike ________

12. bin ________

13. kid ________

14. ripe ________

15. fine ________

16. dive ________

17. stick ________

18. ride ________

19. fix ________

20. dip ________

Name ____________________________________

hike	did	rip	pipe	side
fin	mix	nine	five	pick

Match-Ups

Draw a line from each spelling word to its meaning.

1. rip	to walk in the woods
2. hike	part of a fish
3. pipe	to tear
4. mix	a metal tube
5. fin	to stir

Sentences to Complete

Write a spelling word to complete each sentence.

6. I ______________ the puzzle all by myself.

7. Can I ______________ some flowers for you?

8. I know that ten comes after ______________.

9. I painted one ______________ of the box.

10. I could read when I was ______________ years old.

Name ________________________________

There are six spelling mistakes in the list below. Circle the misspelled words. Write the words correctly on the lines below.

Things I Can Do

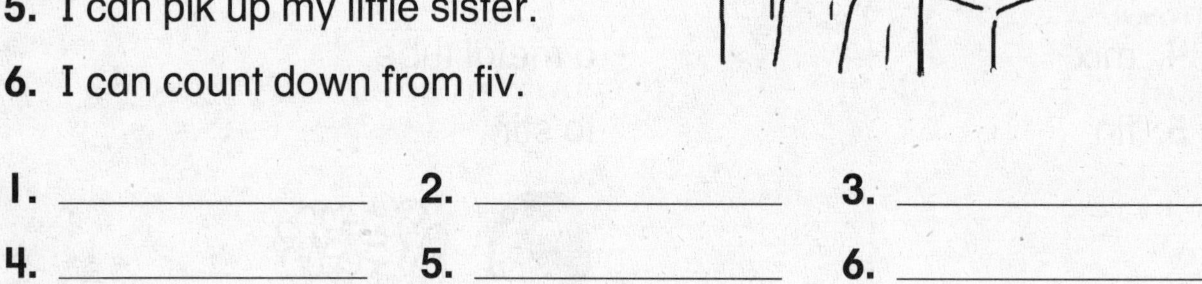

1. I can hik up a big hill.
2. I can mixx red and yellow paint.
3. I can spell the word nien.
4. I can clean up my sied of the room.
5. I can pik up my little sister.
6. I can count down from fiv.

1. ____________ 2. ____________ 3. ____________

4. ____________ 5. ____________ 6. ____________

Writing

Make a list of things you can do all by yourself. Use four words from the spelling list.

____

____

____

____

Name________________________________

Using the Word Study Steps

1. LOOK at the word.
2. SAY the word aloud.
3. STUDY the letters in the word.
4. WRITE the word.
5. CHECK the word.
   Did you spell the word right?
   If not, go back to step 1.

Spelling Words

dog	hope
fox	rope
lock	pot
rose	box
poke	cone

Puzzle

Solve the puzzle. Circle the five hidden spelling words.

w	y	q	f	o	x	p
k	c	o	n	e	s	r
p	o	t	x	h	m	z
r	w	l	r	o	p	e
d	a	c	j	d	o	g

At Home: Review the Word Study Steps with your child as you both go over this week's spelling words.

Name ______________________

dog	lock	poke	rope	box
fox	rose	hope	pot	cone

Word Sort

Look at the spelling words in the box. Write the spelling words that have the short *o* sound.

1. ____________ **2.** ____________ **3.** ____________

4. ____________ **5.** ____________

Write the spelling words that have the long *o* sound.

6. ____________ **7.** ____________ **8.** ____________

9. ____________ **10.** ____________

Misfit Letter

An extra letter has been added to each spelling word below. Draw a line through the letter that does not belong. Write the word correctly on the line.

11. boxx ____________ **12.** roepe ____________

13. doag ____________ **14.** coine ____________

15. hopie ____________ **16.** pooke ____________

17. faox ____________ **18.** locke ____________

19. pout ____________ **20.** roase ____________

Name ____________________

dog	lock	poke	rope	box
fox	rose	hope	pot	cone

Make a Connection

Write a spelling word to complete each pair of sentences.

1. A cat purrs.
   A ______________ barks.
2. The skunk smelled bad.
   The ______________ smelled nice.
3. We made eggs in the pan.
   We made soup in the ______________.
4. The bear was black.
   The ______________ was red.
5. The bag was made of paper.
   The ______________ was made of wood.
6. The string was easy to cut.
   The ______________ was hard to cut.

Sentences to Complete

Write a spelling word to complete each sentence.

7. Put the key in the ______________.
8. Mom put ice cream in the ______________.
9. I ______________ you can come to my party.
10. Do not ______________ me with the stick.

Name ______________________________

There are five spelling mistakes in the paragraph below. Circle the misspelled words. Write the words correctly on the lines below.

There is a new girl in my class. I hoape we can be friends. I will show her where to put her lunch boxe. I will tell her about my doig. I can teach her how to play jump roepe at recess. Then we can have an ice cream coyne after school.

1. ____________ **2.** ____________ **3.** ____________

4. ____________ **5.** ____________

Writing

Write about how you would become friends with a new boy or girl who came to your school. Use five words from the spelling list.

____

____

____

____

____

Name ______________________________

Using the Word Study Steps

1. LOOK at the word.
2. SAY the word aloud.
3. STUDY the letters in the word.
4. WRITE the word.
5. CHECK the word.
   Did you spell the word right?
   If not, go back to step 1.

Spelling Words

sun	bud
duck	bump
cup	cube
dude	fume
rule	rude

Circle the Word

Circle the words with the long *u* sound.

duck	bump	cup	cube	rude
sun	fume	dude	rule	bud

At Home: Review the Word Study Steps with your child as you both go over this week's spelling words.

Name ______________________________

bud	rude	fume	sun	cup
dude	rule	duck	bump	cube

Word Sort

Look at the spelling words in the box. Write each spelling word in the correct column.

Short *u* Words	**Long *u* Words**
1. ______________	6. ______________
2. ______________	7. ______________
3. ______________	8. ______________
4. ______________	9. ______________
5. ______________	10. ______________

Rhyme Time

Write the spelling word that rhymes with each of these words.

11. bun ______________
12. lump ______________
13. mud ______________
14. luck ______________
15. pup ______________

Name____

has	sat	wag	had	bad
fix	six	him	will	if

Match-Ups

Draw a line from each spelling word to its meaning.

1. fume	something you drink from
2. cup	a smell
3. duck	a shape with six sides
4. cube	to hit
5. bump	a bird that swims

Sentences to Complete

Write a spelling word on each line to complete the sentence.

6. It is hot sitting in the ______________.
7. It is not nice to be ______________.
8. There is a pink ______________ on the plant.
9. Please follow the class ______________.
10. Gus is a cool ______________.

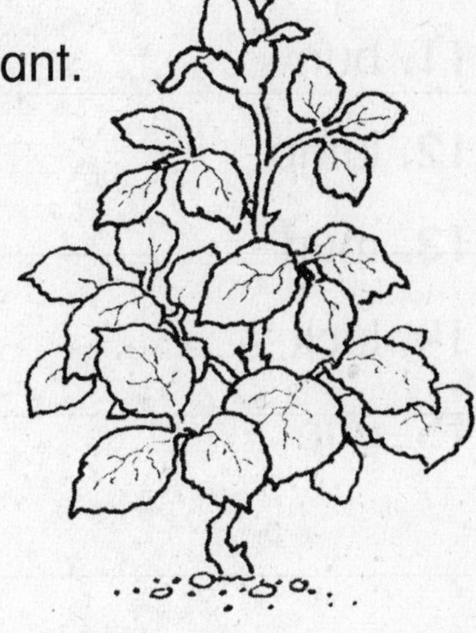

Name ____________________

There are five spelling mistakes in the paragraph below. Circle the misspelled words. Write the words correctly on the lines below.

Dad and I planted seeds today. We put them where they would get lots of suen. I gave each seed a cuip of water. I made a sign with a ruel. The sign said: *Please do not bummp the plants.* I can't wait until we see the first bude.

1. __________ **2.** __________ **3.** __________

4. __________ **5.** __________

Writing

Write about planting or taking care of seeds. Use five spelling words from your list.

____

____

____

____

____

Name ____________________

Spelling Words

slide	task
slips	still
dress	must
drop	crisp
skin	spell

Using the Word Study Steps

1. LOOK at the word.
2. SAY the word aloud.
3. STUDY the letters in the word.
4. WRITE the word.
5. CHECK the word.
   Did you spell the word right?
   If not, go back to step 1.

Word Builder

Be a word builder. Write the missing consonant to finish each spelling word.

1. s ____ ips
2. d ____ ess
3. s ____ ell
4. mus ____
5. d ____ op
6. s ____ in
7. tas ____
8. s ____ ide
9. s ____ ill
10. cris ____

At Home: Review the Word Study Steps with your child as you both go over this week's spelling words.

Name________________________

drop	skin	task	spell	crisp
slide	must	still	dress	slips

Word Sort

Find the spelling words that begin or end with each of the sounds below. Write the words in the correct box.

sl

1. ____________
2. ____________

dr

3. ____________
4. ____________

dr

5. ____________
6. ____________

st

7. ____________
8. ____________

sp

9. ____________
10. ____________

Pattern Smart

Write the spelling words that have the same pattern as *drum*.

11. ____________ 12. ____________

Write the spelling words that have the same pattern as *slap*.

13. ____________ 14. ____________

15. Where do these letters appear? Circle the answer.

at the beginning   at the end

Name ______________________________

dress	task	drop	skin	still
slips	must	crisp	spell	slide

Not the Same

On the line beside each word, write the spelling word that means the opposite.

1. pick up ______________
2. soggy ______________
3. undress ______________
4. moving ______________

Sentences to Complete

Write a spelling word on each line to complete the sentence.

6. Can I go down the ______________?
7. I can ______________ a lot of words.
8. My ______________ is red from the sun.
9. Dad and Mom ______________ pay their bills.
10. The ______________ was to wash the windows.

Name ____________________

There are five spelling mistakes in the diary entry below. Circle the misspelled words. Write the words correctly on the lines below.

May 10 Today I found a stray kitten. I knew I musk take care of her. I took her inside. I think she was so scared that she sat stille. I gave her a dropp of milk. She liked it! She started to lick her fur. Her scin was red. I will need to take her to the vet. Finding who owns her will not be a easy tassk. But I know I will!

1. ____________ 2. ____________ 3. ____________

4. ____________ 5. ____________

Writing

Write about how you would help rescue an animal. Use five spelling words from the list.

____

____

____

____

____

Name ______________________________

Spelling Words

main	jay
wait	pay
sail	stay
tail	hay
train	may

Using the Word Study Steps

1. LOOK at the word.
2. SAY the word aloud.
3. STUDY the letters in the word.
4. WRITE the word.
5. CHECK the word.
   Did you spell the word right?
   If not, go back to step 1.

Find and Circle

Where are the spelling words?

p	m	a	i	n	t	o	h	a	y	i	g
t	a	s	a	i	l	e	k	o	j	a	y
a	y	z	u	p	a	y	w	a	i	t	r
i	f	n	t	r	a	i	n	h	v	a	s
l	m	d	a	j	k	s	t	a	y	i	p

At Home: Review the Word Study Steps with your child as you both go over this week's spelling words.

Name ______________________________

jay	may	wait	sail	train
hay	main	tail	pay	stay

Write the Words

Write the spelling words that have the long *a* sound spelled *ai*.

1. ____________ 2. ____________ 3. ____________

4. ____________ 5. ____________

Write the spelling words that have the long *a* sound spelled *ay*.

6. ____________ 7. ____________ 8. ____________

9. ____________ 10. ____________

New Words

Make a new word from the spelling list by changing the first letter.

11. way – w + j = ____________

12. mail – m + t = ____________

13. day – d + h = ____________

14. gain – g + m = ____________

15. brain – b + t = ____________

Name ______________________________

tail	stay	pay	main	wait
may	sail	hay	jay	train

Sentences to Complete

Write the spelling word on each line to complete the sentence.

1. I have to ______________ a quarter for the milk.
2. My dog wags his ______________ when he is happy.
3. Do you know how to ______________ a boat?
4. Will you ______________ for me to get there?
5. I ______________ not go to the game today.
6. What time does the ______________ come?
7. Grandma will ______________ at our house.
8. Is that a blue ______________ in the tree?
9. There were lots of shops on the ______________ street.
10. There is plenty of ______________ in the barn.

There are five spelling mistakes in the paragraph below. Circle the misspelled words. Write the words correctly on the lines below.

Today my mom fell and hurt her arm. My dad thought she mae need a cast. We all went to the hospital. We needed to see the mayn doctor. We had to waet our turn. Then my mom got an x-ray. She did need a cast. The doctor told her to staiy still. My dad left to paye the bill. At last, we all went home and signed Mom's new cast.

1. ______________ 2. ______________ 3. ______________
4. ______________ 5. ______________

Writing

Write about how you would help someone who got hurt or who was sick. Use five spelling words from your list.

____

____

____

____

____

Name ______________________________

Using the Word Study Steps

1. LOOK at the word.
2. SAY the word aloud.
3. STUDY the letters in the word.
4. WRITE the word.
5. CHECK the word.
   Did you spell the word right?
   If not, go back to step 1.

Spelling Words

light	high
sight	wild
mind	dry
cry	try
tie	lie

X the Words

Put an X on the words with the long *i* sound.

dry	dip	pick	lie	rip
sit	sight	mit	try	mind
wild	tip	high	clip	tie
hill	cry	light	tilt	will

At Home: Review the Word Study Steps with your child as you both go over this week's spelling words.

Name ______________________________

light	lie	try	high	tie
wild	mind	sight	cry	dry

Write the Words

Write the spelling words that have the long *i* sound spelled *i*.

1. __________ **2.** __________

Write the spelling words that have the long *i* sound spelled *ie*.

3. __________ **4.** __________

Write the spelling words that have the long *i* sound spelled *y*.

5. __________ **6.** __________ **7.** __________

Write the spelling words that have the long *i* sound spelled *igh*.

8. __________ **9.** __________ **10.** __________

Misfit Letter

An extra letter has been added to each spelling word below. Draw a line through the letter that does not belong. Write the word correctly on the line.

11. highe __________ **12.** miend __________

13. crye __________ **14.** tyie __________

15. wiled __________

Name ________________________________

light	lie	try	high	tie
wild	mind	sight	cry	dry

Word Meaning

Find the opposite. Draw lines to connect the spelling words to words that mean the opposite.

1. dry — low
2. high — wet
3. wild — heavy
4. light — tame

Sentences to Complete

Write a spelling word on each line to complete the sentence.

5. Can you ______________ your shoes?
6. I will ______________ to help you fix the car.
7. You look sad when you ______________.
8. Your ______________ is what helps you see.
9. Do you ______________ if I sit next to you?
10. You should never tell a ______________.

Name ______________________

There are five spelling mistakes in the report below. Circle the misspelled words. Write the words correctly on the lines below.

Our class took a trip to the zoo. We saw tame animals and wield animals. There was a baby kangaroo. He could jump highe. We did not miend getting splashed by the baby seal pups. We sat in the sun to get driy. We liked the newborn lion cubs the best.The zoo keeper told us not to trye to feed them. Their mother might get mad!

1. ______________ **2.** ______________ **3.** ______________

4. ______________ **5.** ______________

Writing

Write a report about baby or adult animals. Use five words from the spelling list.

____

____

____

____

Name ____________________

Spelling Words

grow	toast
mow	soap
crow	foam
toe	told
goes	most

Using the Word Study Steps

1. LOOK at the word.
2. SAY the word aloud.
3. STUDY the letters in the word.
4. WRITE the word.
5. CHECK the word.
   Did you spell the word right?
   If not, go back to step 1.

Crossword Puzzle

Write the spelling word that best matches each clue. Put the spelling words in the boxes that start with the same number.

ACROSS

2. past tense of *tell*
3. almost all
7. to get bigger
8. what you wash with

DOWN

1. moves
3. to cut grass
4. to heat bread
5. a bird
6. soap bubbles

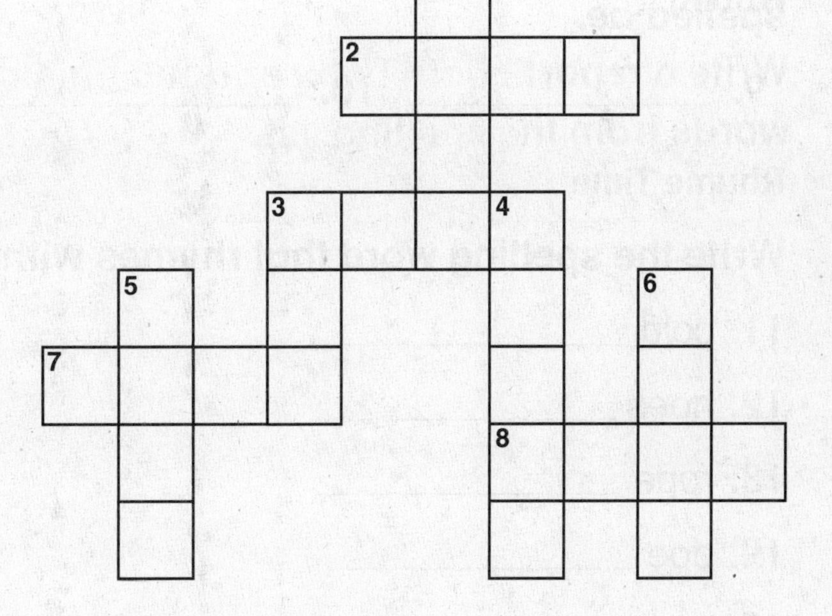

At Home: Review the Word Study Steps with your child as you both go over this week's spelling words.

Name ______________________________

most	mow	goes	toast	foam
grow	told	crow	toe	soap

Write the Words

Write the spelling words that have the long *o* sound spelled *o*.

1. ____________ **2.** ____________

Write the spelling words that have the long *o* sound spelled *oa*.

3. ____________ **4.** ____________ **5.** ____________

Write the spelling words that have the long *o* sound spelled *ow*.

6. ____________ **7.** ____________ **8.** ____________

Write the spelling words that have the long *o* sound spelled *oe*.

9. ____________ **10.** ____________

Rhyme Time

Write the spelling word that rhymes with each of these words.

11. cold ____________

12. hoes ____________

13. rope ____________

14. doe ____________

15. roast ____________

Name ______________________________

toast	soap	mow	goes	crow
grow	toe	told	most	foam

Sentences to Complete

Write a spelling word on each line to complete the sentence.

1. I ______________ my friend to wait for me after class.
2. What color are ______________ trees?
3. Dad will ______________ the grass today.
4. Mom made eggs and ______________.
5. There is a bar of ______________ on the sink.
6. A ______________ is a black bird.
7. Did you ______________ an inch taller?
8. I stubbed my ______________ on the step.
9. Joe ______________ to work at nine.
10. The soap made bubbles and ______________.

Name ____________________

There are five spelling mistakes in the paragraph below. Circle the misspelled words. Write the words correctly on the lines below.

When I groe up I want to play baseball like my brother. He gois to practice every day. He tolde me that he works hard to be a good player. He has never missed a game. Once he even played with a broken tow. What he loves moast about baseball is that it helps keep him in shape.

1. ____________ 2. ____________ 3. ____________

4. ____________ 5. ____________

Writing

What sport or activity keeps you in good shape? Write about it. Use five spelling words from your list.

____

____

____

____

____

Name ____________________

Using the Word Study Steps

1. LOOK at the word.
2. SAY the word aloud.
3. STUDY the letters in the word.
4. WRITE the word.
5. CHECK the word.
   Did you spell the word right?
   If not, go back to step 1.

Spelling Words

eat	leaf
mean	queen
need	seek
baby	pony
he	we

Puzzle

Solve the puzzle. Circle the ten hidden spelling words.

o	e	a	t	y	h	e
p	m	q	u	e	e	n
e	b	a	b	y	w	e
m	e	a	n	b	h	s
e	a	c	p	o	n	y
s	z	l	e	a	f	u
n	e	e	d	h	s	b
a	y	k	s	e	e	k

At Home: Review the Word Study Steps with your child as you both go over this week's spelling words.

Name ______________________________

need	baby	we	queen	eat
leaf	he	mean	seek	pony

Word Sort

Fill in the blanks below with spelling words that match each spelling pattern.

e	*ee*	*ea*	*y*
1. ________	3. ________	6. ________	9. ________
2. ________	4. ________	7. ________	10. ________
	5. ________	8. ________	

New Words

Make a new word from the spelling list by changing the first letter.

11. seed – s + n ____________
12. be – b + w ____________
13. lean – l + m ____________
14. me – m + h ____________
15. peek – p + s ____________

Name ______________________________

pony	he	seek	mean	need
leaf	eat	baby	we	queen

Match-Ups

Draw a line from each spelling word to its meaning.

1. pony	a small horse
2. baby	to take in food
3. leaf	not nice
4. seek	a very young child
5. mean	part of a plant
6. eat	to try to find

Sentences to Complete

Write a spelling word on each line to complete the sentence.

7. Can ______________ all go to the play with you?

8. The king and ______________ wore crowns.

9. How will ______________ find his way home?

10. I ______________ to get some sleep.

Name ______________________________

There are six spelling mistakes in the paragraph below. Circle the misspelled words. Write the words correctly on the lines below.

My grandpa likes to tell stories. He tells me about things I did when I was a little babie. He remembers when wey went to the park together. He says that when I was a baby, I did not eet very much. Now I nead a lot of food to fill me up! Hee tells me about the first time I rode on a ponie. I did not want to get off. I hope my grandpa keeps telling me stories.

1. ______________ 2. ______________ 3. ______________

4. ______________ 5. ______________ 6. ______________

Writing

Write a family story. Use four words from the spelling list. Share and compare your story with a classmate's story.

____

____

____

____

Name ______________________________

Using the Word Study Steps

1. LOOK at the word.
2. SAY the word aloud.
3. STUDY the letters in the word.
4. WRITE the word.
5. CHECK the word.
   Did you spell the word right?
   If not, go back to step 1.

Spelling Words

mule	June
bugle	music
fuse	duke
use	tune
flute	dune

X the Words

Put an X on the words with the long *u* sound.

hut	fuse	bun	bugle	button
tune	rug	dune	use	us
bug	music	luck	duke	mule
flute	much	cup	June	sun

At Home: Review the Word Study Steps with your child as you both go over this week's spelling words.

Name ______________________

flute	tune	dune	use	June
mule	duke	bugle	music	fuse

Word Sort

Look at the spelling words in the box. Write the spelling words that have the long *u* sound spelled *u*.

1. ______________ 2. ______________

Write the spelling words that have the long *u* sound spelled *u_e*.

3. ______________ 4. ______________ 5. ______________

6. ______________ 7. ______________ 8. ______________

9. ______________ 10. ______________

Puzzle

Solve the puzzle. Circle the five hidden spelling words.

e	u	m	u	l	e	z
i	f	l	u	t	e	s
u	s	e	f	n	k	g
e	c	n	t	u	n	e
y	d	u	n	e	l	m

Name ______________________________

flute	tune	dune	use	June
mule	duke	bugle	music	fuse

Match-Ups

Draw a line from each spelling word to its meaning.

1. June	a sand hill
2. flute	the month before July
3. dune	a nobleman
4. mule	an animal like a donkey
5. duke	a wind instrument

Sentences to Complete

Write a spelling word on each line to complete the sentence.

6. I like ______________ class because I get to play the drums.
7. A ______________ blew so all the lights went out.
8. A ______________ is like a trumpet.
9. Can I ______________ your pen to write a note?
10. Dad hums a ______________ when he rocks the baby.

Name ______________________________

There are five spelling mistakes in the list of rules below. Circle the misspelled words. Write the words correctly on the lines below.

Rules for Music Class

1. Do not uise instruments without asking the teacher.
2. Tuune your instrument before class begins.
3. Make sure no one will trip on your buggle.
4. Keep your fluit in the case unless you are playing it.
5. Put all of the musec books in a neat stack before you leave.

1. ____________ 2. ____________ 3. ____________

4. ____________ 5. ____________

Writing

Write about other school rules that are important to follow. Use five spelling words from your list.

____

____

____

____

Name ____________________

Spelling Words	
chest	chill
chase	shape
sheep	thing
think	white
while	wheat

Using the Word Study Steps

1. LOOK at the word.
2. SAY the word aloud.
3. STUDY the letters in the word.
4. WRITE the word.
5. CHECK the word.
   Did you spell the word right?
   If not, go back to step 1.

Find and Circle

Where are the spelling words?

h	c	h	e	s	t	a	w	t	h	i	n	k
c	h	a	s	e	p	s	h	a	p	e	h	s
t	i	c	z	t	s	h	e	e	p	k	m	z
h	l	w	h	i	t	e	a	w	h	i	l	e
x	l	s	h	u	p	y	t	h	i	n	g	e

At Home: Review the Word Study Steps with your child as you both go over this week's spelling words.

Name ____________________

chase	wheat	think	chest	shape
sheep	white	chill	thing	while

Word Sort

Look at the spelling words in the box. Write the spelling words that follow the patterns below.

words with *th*

1. ____________

2. ____________

words with *sh*

3. ____________

4. ____________

words with *wh*

5. ____________

6. ____________

7. ____________

words with *ch*

8. ____________

9. ____________

10. ____________

Sound the Same

Write the spelling words that have the same pattern as *ship*.

11. ____________ 12. ____________

Write the spelling words that have the same pattern as *cheap*.

13. ____________ 14. ____________

15. ____________

16. Circle the letters that spell the pattern in each word you wrote. Where do these letters appear? Circle the answer.

at the beginning in the middle at the end

Name ________________________________

thing	sheep	chill	shape	wheat
chest	while	chase	white	think

Opposites

Draw a line to connect the words that mean the opposite.

1. chill	black
2. white	warm up
3. chase	let go

Sentences to Complete

Write a spelling word on each line to complete the sentence.

4. A circle is a round ________________.
5. Wool comes from ________________.
6. I ________________ we should go home now.
7. Mom reads the map ________________ Dad drives.
8. Is there ________________ in the bread?
9. The toy ________________ was filled with blocks.
10. What ________________ is big, red, and shiny?

Name ______________________________

There are six spelling mistakes in the paragraph. Circle the misspelled words. Write the words correctly on the lines below.

Julie and I were digging in the dirt. Julie found a whiet rock. The shaip of the rock was round. She looked closely at the rock whyle I kept on digging. Julie said, "I thinke this might be a fossil. I see a thinng that looks like a bone." Then a chille ran up my spine. Could this be the fossil of a dinosaur?

1. ______________ 2. ______________ 3. ______________

4. ______________ 5. ______________ 6. ______________

Writing

Write about what it would be like to find the fossil of an animal that lived millions of years ago. Use four spelling words from your list.

____

____

____

____

Name ____

Using the Word Study Steps

1. LOOK at the word.
2. SAY the word aloud.
3. STUDY the letters in the word.
4. WRITE the word.
5. CHECK the word.
   Did you spell the word right?
   If not, go back to step 1.

Spelling Words

each	which
teaching	path
teeth	fishing
wishbone	watch
matching	dish

X the Word

Look at the end part of the spelling words in each row. Put an X on the word that does not belong.

1. watch	matching	fishing
2. each	dish	which
3. fishing	wishbone	matching
4. path	teeth	each
5. teaching	matching	watch

At Home: Review the Word Study Steps with your child as you both go over this week's spelling words.

Name ____________________

which	teaching	each	dish	matching
path	wishbone	teeth	fishing	watch

Pattern Power

Write the spelling words for each of these patterns.

th

1. ____________________
2. ____________________

sh

3. ____________________
4. ____________________
5. ____________________

ch

6. ____________________
7. ____________________
8. ____________________

tch

9. ____________________
10. ____________________

Rhyme Time

Write the spelling word that rhymes with each of these words.

11. peach ____________________
12. bath ____________________
13. fish ____________________
14. rich ____________________
15. reaching ____________________

Name ____________________

teeth	watch	dish	matching	each
which	teaching	path	wishbone	fishing

Sentences to Complete

Write a spelling word on each line to complete the sentence.

1. Trish put her sandwich on a ______________.
2. Hold the ______________ and make a wish.
3. Brush your ______________ three times a day.
4. We rode our bikes on the dirt ______________.
5. I don't know ______________ street to take.
6. What time does your ______________ say?
7. Mitch got ______________ one of us a gift.

Word Building

Add *-ing* to each word. Write the spelling word.

8. teach + ing = ______________
9. fish + ing = ______________
10. match + ing = ______________

Name ____________________

There are six spelling mistakes in the invitation below. Circle the misspelled words. Write the words correctly on the lines below.

Dear Parents,

Please come to our class play. Mrs. Jones is teashing us some new songs. Eatch one of us will have a special part. We are going to wear matshing costumes and hats. We hope you will come and wach us!

The play is in the gym. Follow the red pacth to find your seat. The shows are Friday, Saturday, and Sunday. Whitch show will you come to see?

1. ____________ 2. ____________ 3. ____________

4. ____________ 5. ____________ 6. ____________

Writing

Invite someone to come watch you sing, dance, or perform something special. Use four spelling words from your list.

____

____

____

Name ______________________________

Spelling Words

screen	strict
scream	sprain
scrape	spring
stripe	spruce
struck	strap

Using the Word Study Steps

1. LOOK at the word.
2. SAY the word aloud.
3. STUDY the letters in the word.
4. WRITE the word.
5. CHECK the word.
   Did you spell the word right?
   If not, go back to step 1.

Find and Circle

Where are the spelling words?

s c r a p e f s r j l s t r a p
p t l b q r k c g s p r i n g c
r v r f x s t r u c k t d g y z
u m y i b j y e h k s p r a i n
c s w q c n b a s c r e e n p r
e p d z m t d m z x n d f c k q
x s t r i p e f q w b c h s z p

At Home: Review the Word Study Steps with your child as you both go over this week's spelling words.

Name ____________________

strap	spruce	strict	struck	scream
scrape	spring	sprain	stripe	screen

Word Sort

Find the spelling words that begin with each of the letters below. Write the words on the lines.

scr	*str*	*spr*
1. ________	4. ________	8. ________
2. ________	5. ________	9. ________
3. ________	6. ________	10. ________
	7. ________	

Missing Letter

A letter is missing from each spelling word below. Write the missing letter in the box. Then write the spelling word correctly on the line.

11. spain ☐ ________
12. sream ☐ ________
13. srict ☐ ________
14. spuce ☐ ________
15. sripe ☐ ________

Name ____________________

spring	strap	spruce	strict	screen
scream	scrape	stripe	struck	sprain

Match-Ups

Draw a line from each spelling word to its meaning.

1. scrape	to yell
2. scream	the season after winter
3. sprain	a kind of tree
4. spruce	a leather or cloth band
5. strap	to rub
6. spring	to twist a muscle

Sentences to Complete

Write a spelling word on each line to complete the sentence.

7. The ______________ on the window keeps the bugs out.

8. Dan had a red ______________ on his soccer uniform.

9. Last time I was up to bat I ______________ out.

10. Our principal is very ______________ about school rules.

Name ____

There are six spelling mistakes in the story below. Circle the misspelled words. Write the words correctly on the lines below.

Kate grew up on a dairy farm. Her parents were very scrict. Everyone had to get up at 5:00 in the morning to help with chores. Kate would help milk the cows. Her favorite cow had a black strippe on its front leg.

One sping day it was raining very hard. When Kate went to help feed the animals, she slipped in the mud. She fell and hurt her ankle. The doctor said it was just a scrain and that it would heal quickly. Kate had to stay off of her feet. It struc her then how much she missed being around the animals. When Kate's ankle was better she went horseback riding. She let out a loud skream because she was so happy.

1. ______________ 2. ______________ 3. ______________

4. ______________ 5. ______________ 6. ______________

Writing

Write about how you could help out on a farm. Use four spelling words in your list.

____

____

____

Name ____________________

Using the Word Study Steps

1. LOOK at the word.
2. SAY the word aloud.
3. STUDY the letters in the word.
4. WRITE the word.
5. CHECK the word.
   Did you spell the word right?
   If not, go back to step 1.

Spelling Words	
part	sort
start	storm
park	short
farm	for
dark	horse

Find and Circle

Solve the puzzle. Circle the ten hidden spelling words.

h	o	r	s	e	s	t	a	r	t
p	t	a	h	f	a	r	m	u	d
a	k	s	o	r	t	n	o	s	a
r	p	a	r	t	w	c	f	o	r
k	i	z	t	s	t	o	r	m	k

At Home: Review the Word Study Steps with your child as you both go over this week's spelling words.

Name ______________________

horse	for	part	start	short
farm	park	sort	dark	storm

Word Sort

Look at the spelling words in the box. Match each word to a spelling pattern.

ar	***or***
1. ______	6. ______
2. ______	7. ______
3. ______	8. ______
4. ______	9. ______
5. ______	10. ______

Circle the Word

Circle the words with *or*.

short	dark	farm	part	for
start	sort	horse	storm	park

Name____

dark	farm	storm	for	part
horse	short	start	park	sort

Opposites

Write the spelling word that means the opposite of each word below.

1. stop ________________
2. light ________________
3. long ________________

Sentences to Complete

Write a spelling word on the line to complete the sentence.

4. Sam has the biggest ________________ in the play.
5. Cows and pigs live on a ________________.
6. I have a gift ________________ my best friend.
7. Do you know how to ride a ________________?
8. Sue has to ________________ the socks by color.
9. We heard thunder during the ________________.
10. There are swings and a slide at the ________________.

Name ______________________________

There are six spelling mistakes in the paragraph below. Circle the misspelled words. Write the words correctly on the lines below.

My dog can run in the perk. During a starm, my dog hides under the couch. After dirk, my dog sleeps in bed with me. A herse is too big to sleep in a bed. It lives in a stable on a faerm. A horse likes to run fer miles. It eats a lot of food. I think a dog is easier to take care of than a horse.

1. ____________ 2. ____________ 3. ____________

4. ____________ 5. ____________ 6. ____________

Writing

Write about the needs of two different animals. Use four words from your spelling list.

____

____

____

____

____

Name________________________________

Using the Word Study Steps

1. LOOK at the word.
2. SAY the word aloud.
3. STUDY the letters in the word.
4. WRITE the word.
5. CHECK the word.
   Did you spell the word right?
   If not, go back to step 1.

Spelling Words	
clerk	term
herd	skirt
sir	stir
churn	burst
hurt	turn

Word Builder

Be a word builder. Write the missing vowel to finish each spelling word.

1. c l _____ r k
2. t _____ r n
3. h _____ r t
4. s _____ r
5. t _____ r m
6. h _____ r d
7. c h _____ r n
8. b _____ r s t
9. s t _____ r
10. s k _____ r t

At Home: Review the Word Study Steps with your child as you both go over this week's spelling words.

Name ______________________________

term	hurt	sir	herd	stir
skirt	clerk	churn	burst	turn

Word Sort

Look at the spelling words in the box. Match each word with a spelling pattern.

er **1.** __________ **2.** __________ **3.** __________

ir **4.** __________ **5.** __________ **6.** __________

ur **7.** __________ **8.** __________ **9.** __________

10. __________

Misfit Letter

An extra letter has been added to each spelling word below. Draw a line through the letter that does not belong. Write the word correctly on the line.

11. stier __________

12. hierd __________

13. clierk __________

14. huert __________

15. cheurn __________

Name ______________________________

churn	burst	clerk	skirt	term
sir	hurt	turn	stir	herd

Sentences to Complete

Write a spelling word on each line to complete the sentence.

1. The ______________ in the store waited on us.
2. Trish wore a ______________ and a sweater to the dance.
3. The first school ______________ is over in November.
4. You can call the man ______________ to be polite.
5. Please take your ______________ in the game.

Word Meaning

Say it another way. Draw a line from each spelling word to the word or words that mean almost the same.

6. burst	mix
7. stir	pop
8. herd	stir milk
9. hurt	wounded
10. churn	large group

Name ___________________________________

There are five spelling mistakes in the paragraph below. Circle the misspelled words. Write the words correctly on the lines below.

Animals have many different needs. Jason knows this from helping on his father's ranch. It is Jason's tirn to help with the cattle. There is a large hurd. They all need to stay together. Jason also needs to make sure that none of the cattle gets hert. Jason needs to stur a special medicine into the food of one cow that is sick. This is a big job. It takes a berst of energy for Jason to take care of the whole herd.

1. ____________ 2. ____________ 3. ____________

4. ____________ 5. ____________

Writing

Write a paragraph about the needs of one of your favorite animals. Use five words from your spelling list.

____

____

____

____

Name ____________________

Using the Word Study Steps

1. LOOK at the word.
2. SAY the word aloud.
3. STUDY the letters in the word.
4. WRITE the word.
5. CHECK the word.
   Did you spell the word right?
   If not, go back to step 1.

Spelling Words	
shook	stood
hook	brook
crook	foot
soot	could
should	would

Find and Circle

Where are the spelling words?

w	o	u	l	d	q	w	s	r	t	y
m	s	h	o	o	k	n	t	b	v	c
w	q	k	s	d	f	g	s	h	j	k
c	o	u	l	d	b	n	t	m	h	w
r	b	r	o	o	k	t	o	q	o	p
o	s	o	o	t	h	j	o	k	o	z
o	x	s	h	o	u	l	d	c	k	b
k	b	n	m	q	w	f	o	o	t	r

At Home: Review the Word Study Steps with your child as you both go over this week's spelling words.

Name ____________________

would	shook	should	hook	could
soot	brook	foot	crook	stood

Word Sort

Look at the spelling words in the box. Match the spelling word with the spelling pattern and write the word.

oot 1. ____________ 2. ____________

ook 3. ____________ 4. ____________

5. ____________ 6. ____________

ood 7. ____________

ould 8. ____________ 9. ____________

10. ____________

Pattern Smart

Write the spelling words that have the same pattern as *book*.

11. ____________ 12. ____________

13. ____________ 14. ____________

Write the spelling word that has the same pattern as *hood*.

15. ____________

Name ________________________________

hook	brook	would	should	shook
stood	crook	foot	could	soot

Sentences to Complete

Write a spelling word on each line to complete the sentence.

1. The ______________ stole a watch from the shop.
2. An inch is smaller than a ______________.
3. Hang your coat up on the ______________.
4. We saw ducks swimming in the ______________.
5. I ______________ up so long my feet hurt.
6. He ______________ not be able to play in the game.
7. The little boy ______________ with fear.
8. Mom knew I ______________ pick her up at the mall.
9. There was ______________ in the fireplace.
10. You ______________ know the answer to this question.

Name ____________________

There are six spelling mistakes in the paragraph below. Circle the misspelled words. Write the words correctly on the lines below.

Our class stud by the brouk. It was littered with trash. We knew we shood do something. We got some garbage bags and gloves. We started picking up the trash. Jan's fut almost slipped into the brook. We had to be careful. Someone cuold get hurt. But we knew everyone woold be very happy that we took care of the brook.

1. ____________ 2. ____________ 3. ____________

4. ____________ 5. ____________ 6. ____________

Writing

Write about cleaning up something to make Earth a better place. Use four spelling words from your list.

____

____

____

____

____

Name____

Using the Word Study Steps

1. LOOK at the word.
2. SAY the word aloud.
3. STUDY the letters in the word.
4. WRITE the word.
5. CHECK the word.
   Did you spell the word right?
   If not, go back to step 1.

Spelling Words	
root	glue
boot	flew
suit	new
fruit	shoe
clue	canoe

Find and Circle

Circle the ten hidden spelling words.

b	r	o	o	t	k	n	e	w	q
o	p	c	a	n	o	e	l	j	s
o	g	l	f	r	u	i	t	o	u
t	w	u	s	h	g	l	u	e	i
f	l	e	w	g	s	h	o	e	t

At Home: Review the Word Study Steps with your child as you both go over this week's spelling words.

Name ____________________________________

suit	shoe	root	clue	fruit
glue	flew	canoe	new	boot

Word Sort

Look at the spelling words in the box. Write the spelling words that match each spelling pattern.

oo

1. ________________
2. ________________

ue

3. ________________
4. ________________

ui

5. ________________
6. ________________

ew

7. ________________
8. ________________

oe

9. ________________
10. ________________

Rhyme Around

Write the spelling word that completes each rhyme.

11. Dad has a funny suit. The pattern on it is made of ________________.
12. The toy plane flew because it was brand ________________.
13. We were riding in the canoe when I lost my right ________________.
14. I will give you a clue. This sticky stuff smells like ________________.
15. I was digging up the root when I got mud on my ________________.

Name ______________________________

canoe	boot	fruit	glue	new
root	shoe	clue	suit	flew

Match-Ups

Draw a line from each spelling word to its meaning.

1. root	to make stick
2. glue	part of a plant
3. clue	a set of clothes
4. new	a small boat
5. canoe	recently grown or made
6. suit	a hint

Sentences to Complete

Write a spelling word on each line to complete the sentence.

7. My cowboy ______________ goes up to my knee.

8. I ate a piece of ______________ for lunch.

9. Which ______________ needs a new lace?

10. Mom ______________ to Texas to see her brother.

Name ______________________________

There are five spelling mistakes in the paragraph below. Circle the misspelled words. Write the words correctly on the lines below.

Dave had one buit on when he saw the weather report. A nue cold front was on its way. There was going to be a big winter storm. The big gray clouds were one clew that snow would start falling soon. Dave rushed to the airport. Somehow the pilot floo the plane and landed it before the storm began. Dave saw his friend get off the plane in a sute. Dave gave him a heavy winter coat and gloves for his cold visit to Chicago.

1. ____________ **2.** ____________ **3.** ____________

4. ____________ **5.** ____________

Writing

Write about a big storm.
Use five words from your spelling list.

____

____

____

____

____

Name ____________________

Using the Word Study Steps

1. LOOK at the word.
2. SAY the word aloud.
3. STUDY the letters in the word.
4. WRITE the word.
5. CHECK the word.
   Did you spell the word right?
   If not, go back to step 1.

Spelling Words	
pause	jaw
draw	fawn
launch	hawk
law	raw
fault	crawl

X the Word

Put an X on the word in each row that has a different vowel sound.

1. crawl lamb law
2. wait pause fault
3. draw fawn band
4. raw jaw jam
5. lunch launch hawk

At Home: Review the Word Study Steps with your child as you both go over this week's spelling words.

Name ______________________________

launch	draw	hawk	fawn	pause
law	crawl	fault	raw	jaw

Word Sort

Look at the spelling words in the box. Write the spelling words that have the *au* pattern.

1. __________ **2.** __________ **3.** __________

Write the spelling words that have the *aw* pattern.

4. __________ **5.** __________ **6.** __________

7. __________ **8.** __________ **9.** __________

10. __________

Missing Letter

A letter is missing from each spelling word below. Write the missing letter in the box. Then write the spelling word correctly on the line.

11. pase ☐ __________

12. hak ☐ __________

13. lanch ☐ __________

14. cral ☐ __________

15. falt ☐ __________

Name ____________________

jaw	crawl	fawn	launch	raw
pause	draw	law	fault	hawk

Make a Connection

Write a spelling word to complete each pair of sentences.

1. A child can run. A baby can ____________.
2. A bee is one kind of insect. A ____________ is one kind of bird.
3. I like to paint. You like to ____________.
4. A baby cow is called a calf. A baby deer is called a ____________.
5. Our fingers are part of our hand. Our ____________ is part of our mouth.
6. You need to cook the meat. But carrots you can eat ____________.

Sentences to Complete

Write a spelling word on each line to complete the sentence.

7. They will ____________ the rocket at noon.
8. Is it your ____________ that the vase broke?
9. Wearing your seat belt is a ____________.
10. Stop or ____________ after you read the first page.

Name ______________________________

There are five spelling mistakes in the paragraph below. Circle the misspelled words. Write the words correctly on the lines below.

It is very cold and windy in the Arctic. You're likely to see a baby polar bear living there, but not a little faun. You might also spot a snowy owl, but not a hauk. The Arctic is just too cold for some animals! Animals that have a thick coat of fur can crauwl, jump, or play in the snow. You might pawse and watch a reindeer or moose make tracks in the snow.

What other Arctic animals can you think of? Try to drauw them!

1. ______________ 2. ______________ 3. ______________

4. ______________ 5. ______________

Writing

Write about one or more animals that can survive in the Arctic. Use five spelling words from your list.

____

____

____

____

Name ____

Spelling Words

clown	round
growl	loud
howl	cloud
brown	house
crown	sound

Using the Word Study Steps

1. LOOK at the word.
2. SAY the word aloud.
3. STUDY the letters in the word.
4. WRITE the word.
5. CHECK the word.
   Did you spell the word right?
   If not, go back to step 1.

Puzzle

Solve the puzzle. Circle all the hidden spelling words.

h	r	o	u	n	d	s	h	o	w	l
o	k	v	a	c	r	o	w	n	l	o
u	b	s	d	w	q	u	m	r	y	u
s	i	b	r	o	w	n	z	s	e	d
e	c	l	o	w	n	d	u	y	r	t
g	r	o	w	l	p	c	l	o	u	d

At Home: Review the Word Study Steps with your child as you both go over this week's spelling words.

Name ____________________

Spelling

Words with Diphthong *ou*: *ow* and *ou*: Word Sort

clown	round	crown	loud	cloud
sound	house	brown	growl	howl

Word Sort

Look at the spelling words in the box. Fill in the blanks below with spelling words that match each spelling pattern.

ow	***ou***
1. ____________	6. ____________
2. ____________	7. ____________
3. ____________	8. ____________
4. ____________	9. ____________
5. ____________	10. ____________

Rhyme Time

Write the spelling words that rhyme with each of these words.

11. pound

____

____

12. mouse

____

13. owl

____

____

Name ______________________________

clown	round	crown	loud	cloud
sound	house	brown	growl	howl

Match-Ups

Draw a line from each spelling word to its meaning.

1. clown	shaped like a ball
2. brown	a person who makes you laugh
3. round	a color
4. crown	a building to live in
5. loud	something worn by a king or queen
6. house	noisy

Sentences to Complete

Write a spelling word on the line to complete each sentence.

7. Will the dog ______________ at a stranger?
8. The ______________ in the sky was fluffy and white.
9. There was a loud ______________ when the alarm went off.
10. I think I heard a coyote ______________.

Name ____________________

Spelling
Words with Diphthong *ou*: *ow* and *ou*: Proofreading

There are six spelling mistakes in the report below. Circle the misspelled words. Write the words correctly on the lines below.

A desert is a hot, dry place. It may look broun because few green plants can survive there. Some animals can and do live in the desert. You may hear a lowd soond at night. What is it? It might be the houl of a coyote. Or it might be the groowl of a dingo. Dingoes are like dogs. Some dangerous animals live in the desert, too. If you see one of them, go back into your howse!

1. ________ **2.** ________ **3.** ________

4. ________ **5.** ________ **6.** ________

Writing

Write a short report about animals that live in the desert. Use four of the spelling words in your report.

____

____

____

Name ____________________

Using the Word Study Steps

1. LOOK at the word.
2. SAY the word aloud.
3. STUDY the letters in the word.
4. WRITE the word.
5. CHECK the word.
   Did you spell the word right?
   If not, go back to step 1.

Spelling Words

soil	oil
broil	toy
moist	joy
point	soy
boil	royal

X the Word

Find two words in each row with the same vowel sound and spelling pattern. Cross out the other word that does not belong.

1. royal	crawl	soy
2. soil	moist	most
3. brown	broil	oil
4. joy	job	toy
5. boil	point	paint

At Home: Review the Word Study Steps with your child as you both go over this week's spelling words.

Name ____________________

boil	moist	joy	toy	point
broil	soil	soy	royal	oil

Word Sort

Look at the spelling words in the box. Write the spelling words that have the *oi* pattern.

1. ________ 2. ________ 3. ________

4. ________ 5. ________ 6. ________

Write the spelling words that have the *oy* pattern.

7. ________ 8. ________ 9. ________

10. ________

Missing Letter

A letter is missing from each spelling word below. Write the missing letter in the box. Then write the spelling word correctly on the line.

11. brol ☐ ________

12. roal ☐ ________

13. moit ☐ ________

14. pont ☐ ________

15. sil ☐ ________

Name ______________________________

boil	moist	joy	toy	point
broil	soil	soy	royal	oil

Sentences to Complete

Write a spelling word on each line to complete the sentence.

1. Mom fried the fish in ______________ .
2. The new baby brought much ______________ to her family.
3. I can ______________ the meat in the oven.
4. Water made the towel feel ______________.
5. The ______________ family sat on their thrones.

Definitions

Write the spelling word for each definition.

6. An object that children play with ______________
7. A small mark or dot used in writing ______________
8. Dirt that plants grow in ______________
9. To heat water until it bubbles ______________
10. A sauce used on foods ______________

Name ____________________

There are six spelling mistakes in the paragraph below. Circle the misspelled words. Write the words correctly on the lines below.

Once I was in a play about a king and queen. The stage was a rouyal castle. The queen did not cook. She had her servant boel water for her tea. The king was very funny. He put soey sauce on everything he ate! I played the king and queen's child. I brought them great joiy. My favorite toiy in the castle was a nutcracker. The nutcracker squeaked when I used it. I learned how to oel it so it did not make any noise. The play was fun to be in!

1. ____________ 2. ____________ 3. ____________

4. ____________ 5. ____________ 6. ____________

Writing

Write about acting in a play. Use four or five of your spelling words. Circle the spelling words you use.

____

____

____

____

Name ______________________________

Spelling Words

alone	agree
ago	above
again	awake
away	idea
alike	comma

Using the Word Study Steps

1. LOOK at the word.
2. SAY the word aloud.
3. STUDY the letters in the word.
4. WRITE the word.
5. CHECK the word.
   Did you spell the word right?
   If not, go back to step 1.

Find and Circle

Circle the ten hidden spelling words in the puzzle.

z	a	w	a	k	e	m	a	k	o
a	g	o	w	d	s	n	l	e	c
b	a	p	a	c	v	k	o	b	o
o	i	u	y	a	d	s	n	p	m
v	n	w	t	a	g	r	e	e	m
e	a	l	i	k	e	i	d	e	a

At Home: Review the Word Study Steps with your child as you both go over this week's spelling words.

Name_______________________________________

alone	comma	alike	awake	idea
ago	again	away	agree	above

Word Sort

Look at the spelling words in the box. Write the spelling words that have schwa at the beginning.

1. ________________ **2.** ________________

3. ________________ **4.** ________________

5. ________________ **6.** ________________

7. ________________ **8.** ________________

Write the spelling words that have schwa at the end.

9. ________________ **10.** ________________

Questions and Answers

Write the spelling word that answers each question.

11. What word is the opposite of ***below***? ________________

12. What word describes two things that are the same?

____

13. What word is a punctuation mark? ________________

14. What word means "to have a thought"? ________________

15. What word means "you have the same opinion"? ________________

Name ______________________

alone	comma	alike	awake	idea
ago	again	away	agree	above

Sentences to Complete

Write a spelling word on each line to complete the sentence.

1. The twins dress ____________ every day.
2. We played the game over and over ____________.
3. There is a ____________ in the last sentence.
4. I ____________ with your decision to stay home.
5. The shelf is up ____________ the table.
6. Are you ____________ or sleeping?
7. Whose ____________ was it to clean up the basement?
8. Sam moved far ____________ last year.
9. When you are by yourself, you are ____________.
10. Long ____________ people did not have cars to drive.

Name ______________________________

There are six spelling mistakes in the paragraph below. Circle the misspelled words. Write the words correctly on the lines below.

Long aego there was an explorer named Christopher Columbus. Columbus had the ideea that he would discover new lands. Columbus sailed eway on several voyages. He did not sail ulone. He had a crew on each of his ships. None of his trips was exactly elike. He did not travel to the same place over and over aigain. He discovered many new places in the Caribbean and South America.

1. ____________ 2. ____________ 3. ____________

4. ____________ 5. ____________ 6. ____________

Writing

Write about an explorer who came to America. Use four words from your spelling list.

____

____

____

____

Name ____________________

Using the Word Study Steps

1. LOOK at the word.
2. SAY the word aloud.
3. STUDY the letters in the word.
4. WRITE the word.
5. CHECK the word.
   Did you spell the word right?
   If not, go back to step 1.

Spelling Words	
knee	wrist
knife	wren
knot	thumb
gnaw	lamb
sign	debt

X the Word

Look at the spelling words in each row. Find two words in each row with the same silent letter. Cross out the other word that does not belong.

1.	lamb	thumb	wren
2.	knee	wrist	knot
3.	gnaw	sign	lamb
4.	wren	debt	thumb
5.	sign	knife	knee

At Home: Review the Word Study Steps with your child as you both go over this week's spelling words.

Name ____________________

wrist	gnaw	debt	knife	thumb
knee	sign	wren	lamb	knot

Word Sort

Look at the spelling words in the box. Match each word to a spelling pattern. Write the spelling words on the lines below.

Silent *w*

1. ____________
2. ____________

Silent *k*

3. ____________
4. ____________
5. ____________

Silent *g*

6. ____________
7. ____________

Silent *b*

8. ____________
9. ____________
10. ____________

Missing Letter

The silent letter is missing from each spelling word below. Write the missing letter in the box. Then write the spelling word correctly on the line.

11. nee ☐ ____________
12. det ☐ ____________
13. rist ☐ ____________
14. naw ☐ ____________
15. lam ☐ ____________

Name ______________________________

wren	knot	lamb	knee	debt
knife	gnaw	wrist	thumb	sign

Match-Ups

Draw a line from each spelling word to its meaning.

1. lamb	a part of the arm
2. wrist	a baby sheep
3. knife	a part of the leg
4. knee	a bird
5. wren	a cutting blade

Sentences to Complete

Write a spelling word on each line to complete the sentence.

6. Your shoelace has a big ______________.

7. The ______________ on the door tells visitors where to go.

8. Your ______________ is one of your five fingers.

9. If you owe money, you are in ______________.

10. The beaver will ______________ on the tree bark.

Name ______________________________

Spelling

Words with Silent Letters: *gn, kn, wr, mb*: Proofreading

There are five spelling mistakes in the paragraph below. Circle the misspelled words. Write the words correctly on the lines below.

Grandpa and I planted a garden. Grandpa's rist hurt, so I dug the holes and dropped in the seeds. I had one nee on the ground as I covered the seeds with dirt. Then we made a signe for each vegetable we planted. Grandpa says that I have a green thum. When we finished, we saw a ren flying by. Now Grandpa and I are going to make a scarecrow to keep the birds away!

1. ____________ 2. ____________ 3. ____________

4. ____________ 5. ____________

Writing

Write about planting and taking care of a garden. Use five spelling words. Circle the spelling words you use.

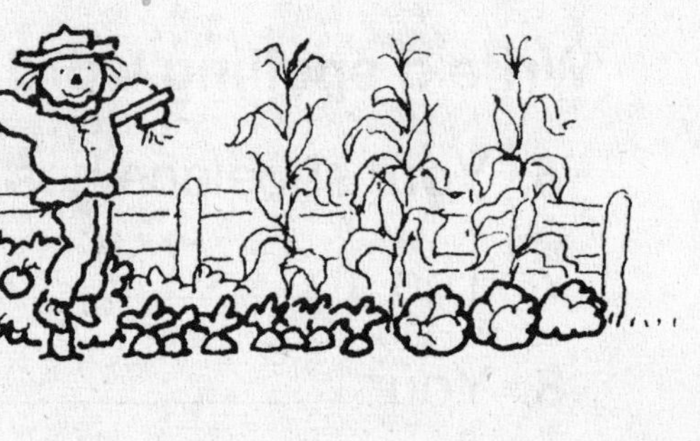

____

____

____

____

Name ____________________

Spelling Words

camp	wagon
cave	gift
cent	gym
face	gem
girl	germ

Using the Word Study Steps

1. LOOK at the word.
2. SAY the word aloud.
3. STUDY the letters in the word.
4. WRITE the word.
5. CHECK the word.
   Did you spell the word right?
   If not, go back to step 1.

Find and Circle

Where are the spelling words?

c	a	m	p	x	g	e	r	m	z
e	q	w	r	t	i	y	p	s	d
n	f	g	h	j	f	g	l	z	x
t	c	v	n	m	t	y	r	w	t
y	p	c	a	v	e	m	d	a	g
h	j	k	l	z	x	c	f	g	b
n	w	g	i	r	l	c	a	o	z
m	e	l	k	j	h	g	c	n	d
s	g	e	m	p	y	t	e	r	w

At Home: Review the Word Study Steps with your child as you both go over this week's spelling words.

Name ______________________________

cave	cent	gift	gym	face
camp	girl	gem	wagon	germ

Word Sort

Look at the spelling words in the box. Write the spelling words that have the sound of soft *c*.

1. ____________ 2. ____________

Write the spelling words that have the sound of hard *c*.

3. ____________ 4. ____________

Write the spelling words that have the sound of soft *g*.

5. ____________ 6. ____________ 7. ____________

Write the spelling words that have the sound of hard *g*.

8. ____________ 9. ____________ 10. ____________

Sounds the Same

If *c* makes the same sound in both words, circle *yes*.
If *c* does not make the same sound in both words, circle *no*.

11. cave, face **yes** **no**

12. camp, cave **yes** **no**

13. face, cent **yes** **no**

Name ______________________________

wagon	gift	camp	cent	gym
face	germ	cave	girl	gem

Match-Ups

Draw a line from each spelling word to its meaning.

1. wagon	the front of the head
2. gift	a cart used by children
3. face	a place where there are tents
4. gym	a present
5. camp	a precious stone
6. gem	a room for games and sports

Sentences to Complete

Write a spelling word on each line to complete the sentence.

7. Bats were flying inside the dark ____________.

8. One ____________ is the same as one penny.

9. Sandy is the name of a ____________ in my class.

10. The ____________ made the little boy sick.

Name ____________________________________

There are five spelling mistakes in the paragraph below. Circle the misspelled words. Write the words correctly on the lines below.

At kamp we learned a lot about the sun and the moon. Why is it dark inside a kave? It is dark because there is no sunlight. The sun is like a jem that brightly shines during the day. At night, it is the moon that shines on your fase. The moon and stars light up the night sky. Sometimes the moon looks round like a one scent coin. Other times you can only see a sliver of the moon. Look up in the sky now. What do you see?

1. ____________ **2.** ____________ **3.** ____________

4. ____________ **5.** ____________

Writing

Write about the day sky and the night sky. Use five spelling words from your list.

____

____

____

____

Name ____________________

Using the Word Study Steps

1. LOOK at the word.
2. SAY the word aloud.
3. STUDY the letters in the word.
4. WRITE the word.
5. CHECK the word.
   Did you spell the word right?
   If not, go back to step 1.

Spelling Words

cage	barge
page	bulge
judge	change
lodge	range
large	hinge

Puzzle

Solve the puzzle. Circle the hidden spelling words.

a	h	i	n	g	e	k	c	a	g	e	n
l	a	r	g	e	s	o	h	p	t	i	r
o	b	e	f	u	t	b	a	r	g	e	a
d	j	u	d	g	e	c	n	z	s	p	n
g	z	t	n	b	u	l	g	e	d	x	g
e	f	x	u	c	n	z	e	p	a	g	e

At Home: Review the Word Study Steps with your child as you both go over this week's spelling words.

Name ______________________________

large	bulge	range	cage	lodge
page	change	judge	hinge	barge

Word Sort

Look at the spelling words in the box. Write the spelling words that end with the spelling pattern *dge*.

1. ______________ 2. ______________

Write the spelling words that end with the spelling pattern *rge*.

3. ______________ 4. ______________

Write the spelling words that end with the spelling pattern *nge*.

5. ______________ 6. ______________ 7. ______________

Write the spelling word that ends with the spelling pattern *lge*.

8. ______________

Write the remaining two words that end with the spelling pattern *ge*.

9. ______________ 10. ______________

Rhyme Time

Write the spelling words that rhyme with each of these words.

rage 11. ______________ 12. ______________

strange 13. ______________ 14. ______________

budge 15. ______________

Name ____

large	bulge	range	cage	lodge
page	change	judge	hinge	barge

Sentences to Complete

Write a spelling word on each line to complete the sentence.

1. Jose's pet rabbit lives in a ______________.
2. The ______________ on the door needs to be oiled.
3. Turn to the last ______________ of your book.
4. Do you think Sam will ______________ his mind?
5. The ______________ box did not fit in the closet.
6. The ______________ banged her gavel in court.
7. People live in the ski ______________ all winter.
8. Your stomach might ______________ if you eat too much.
9. The ______________ floated down the river.
10. The age ______________ is between six and ten years.

Name ______________________

There are five spelling mistakes in the paragraph below. Circle the misspelled words. Write the words correctly on the lines below.

Last Sunday we had a surprise party for Grandma's birthday. The party was at Grandpa's lodje in a largge room. We had to chanje how the tables were arranged so everyone had a place to sit. The time rangge for the party was from four o'clock to eight o'clock, but everyone came early to yell "Surprise!" Grandma was so happy to see her friends. Even the juge who lives next door came. Grandma can't wait until her next birthday!

1. ______________ **2.** ______________ **3.** ______________

4. ______________ **5.** ______________

Writing

Write about a special family celebration. Use five words from the spelling list.

____

____

____

____

Name ____________________

Using the Word Study Steps

1. LOOK at the word.
2. SAY the word aloud.
3. STUDY the letters in the word.
4. WRITE the word.
5. CHECK the word.
   Did you spell the word right?
   If not, go back to step 1.

Spelling Words	
star	dare
shark	hair
care	pair
stare	chair
rare	fair

X the Word

Find two words in each row with the same vowel sound and spelling pattern. Cross out the other word that does not belong.

1. pair	chair	charm
2. star	stamp	shark
3. card	care	rare
4. dare	storm	stare
5. hair	fair	farm

At Home: Review the Word Study Steps with your child as you both go over this week's spelling words.

Name ______________________

fair	care	star	hair	pair
shark	stare	dare	rare	chair

Word Sort

Look at the spelling words in the box. Match each word with a spelling pattern.

ar **1.** ______________ **2.** ______________

are **3.** ______________ **4.** ______________

5. ______________ **6.** ______________

air **7.** ______________ **8.** ______________

9. ______________ **10.** ______________

Misfit Letter

An extra letter has been added to each spelling word below. Draw a line through the letter that does not belong. Write the word correctly on the line.

11. caire ______________ **12.** faire ______________

13. paire ______________ **14.** shairk ______________

15. chaier ______________

Name ______________________________

fair	care	star	hair	pair
shark	stare	dare	rare	chair

Definitions

Write the spelling word for each definition.

1. A piece of furniture that you sit on ____________
2. Not common ____________
3. A large fish ____________
4. An object seen in the night sky ____________
5. To look at something with eyes open wide ____________
6. Two similar things used together ____________
7. A festival or carnival ____________

Sentences to Complete

Write a spelling word on each line to complete the sentence.

8. Wash your ____________ in the shower.
9. I ____________ you to go to school dressed like a monkey.
10. Pat takes ____________ of her new puppy.

Name ______________________________

There are six spelling mistakes in the paragraph below. Circle the misspelled words. Write the words correctly on the lines below.

Everyone in our class is writing a book. Mark's book is about a shairk. Jeff's book is about a shooting starr. My book is about how to take cair of your haire. After we write our books, we will illustrate them. Then we will sit in an author's chare and read our stories to each other. We might also have a book faire so the entire school can read our books.

1. ______________ 2. ______________ 3. ______________

4. ______________ 5. ______________ 6. ______________

Writing

Be an author! Write a story about something you know about or enjoy doing. Use four words from your spelling list.

____

____

____

____

Name ______________________________

Using the Word Study Steps

1. LOOK at the word.
2. SAY the word aloud.
3. STUDY the letters in the word.
4. WRITE the word.
5. CHECK the word.
   Did you spell the word right?
   If not, go back to step 1.

Spelling Words

near	queer
dear	verb
ear	perch
deer	here
steer	where

Find and Circle

Where are the spelling words?

w	s	t	e	e	r	v	o	e	a	r
h	a	w	q	u	e	e	r	l	d	c
e	m	b	o	p	e	r	c	h	e	t
r	x	d	e	e	r	b	h	f	a	s
e	n	e	a	r	l	o	h	e	r	e

At Home: Review the Word Study Steps with your child as you both go over this week's spelling words.

near	where	deer	verb	perch
ear	steer	here	dear	queer

Word Sort

Look at the spelling words in the box. Write the spelling words that have the *er* spelling pattern.

1. ______________ 2. ______________

Write the spelling words that have the *eer* spelling pattern.

3. ______________ 4. ______________ 5. ______________

Write the spelling words that have the *ere* spelling pattern.

6. ______________ 7. ______________

Write the spelling words that have the *ear* spelling pattern.

8. ______________ 9. ______________ 10. ______________

Find the Pattern

Read each group of words. Circle the word that does not fit the pattern.

11. near, ear, deer
12. here, verb, where
13. perch, steer, queer
14. where, dear, near
15. deer, perch, verb

Name ____________________

near	where	deer	verb	perch
ear	steer	here	dear	queer

Match-Ups

Draw a line from each spelling word to its meaning.

1. deer	the part of the body used for listening
2. ear	close by
3. near	an animal with antlers
4. queer	to guide or direct
5. perch	odd or strange
6. steer	a small fish

Sentences to Complete

Write a spelling word on each line to complete the sentence.

7. Please come ____________ so I can show you my fish.
8. I wonder ____________ we are going today in the boat.
9. The word *run* is an example of a ____________.
10. You are a ____________ to take such good care of me.

Name ____________________________________

There are six spelling mistakes in the paragraph below. Circle the misspelled words. Write the words correctly on the lines below.

My mom is an artist. She has a studio wheer she works every day. We live nere the woods so my mom draws lots of animals. One day she drew a der that she saw right outside her window. A fisherman lives next door, so my mom painted a picture of a peerch for him. My mom painted a picture of a cool race car for me. I have it hanging heer in my bedroom. I love that picture. It is very deere to me.

1. ______________ **2.** ______________ **3.** ______________

4. ______________ **5.** ______________ **6.** ______________

Writing

Write about what you would draw or paint if you were an artist. Use four words from your spelling list.

____

____

____

Name ______________________________

north	port	store	roar	board
oar	more	wore	tore	fort

Definitions

Write the spelling word for each definition.

1. Past tense of *tear* ____________
2. A paddle used to row a boat ____________
3. A loud rumbling sound ____________
4. A harbor ____________
5. The direction opposite of south ____________
6. A place where things are sold ____________
7. An army post ____________

Sentences to Complete

Write a spelling word on each line to complete the sentence.

8. I need ____________ glue to finish my art project.
9. Sue ____________ her red dress to the party.
10. The wooden ____________ had nails pounded into it.

Name ______________________________

There are six spelling mistakes in the paragraph below. Circle the misspelled words. Write the words correctly on the lines below.

Inventors have dreamed up many new things over the years. There are moar inventions today than ever before. Someone invented an oare to row a boat. Someone else invented a bord game called checkers. Long ago, if you woore your jeans and toare them, you would need to mend them by hand. Today you can fix them on a sewing machine. Or you can drive to a big department stoore and buy a new pair. What other inventions can you think of?

1. ______________ 2. ______________

3. ______________ 4. ______________

5. ______________ 6. ______________

Writing

Write about your own idea for an invention. Use four words from the spelling list.

____

____

____

Name ________________________________

Using the Word Study Steps

1. LOOK at the word.
2. SAY the word aloud.
3. STUDY the letters in the word.
4. WRITE the word.
5. CHECK the word.
   Did you spell the word right?
   If not, go back to step 1.

Spelling Words

fire	flier
wire	crier
hire	sure
tire	cure
drier	pure

Puzzle

Solve the puzzle. Circle all of the hidden spelling words.

w	d	r	i	e	r	f	p	u	r	e
i	x	o	t	p	u	l	y	f	d	s
r	c	r	i	e	r	i	l	n	d	f
e	u	b	r	l	a	e	s	w	r	i
h	r	p	e	s	u	r	e	l	v	r
k	e	w	z	g	q	s	h	i	r	e

At Home: Review the Word Study Steps with your child as you both go over this week's spelling words.

Name ______________________________

wire	sure	pure	crier	fire
flier	cure	tire	hire	drier

Word Sort

Look at the spelling words in the box. Write the spelling words that have the *ire* pattern.

1. ______________ 2. ______________

3. ______________ 4. ______________

Write the spelling words that have the *ier* pattern.

5. ______________ 6. ______________

7. ______________

Write the spelling words that have the *ure* pattern.

8. ______________ 9. ______________

10. ______________

Match Patterns

If the spelling words in each row have the same pattern, circle *yes*. If they do not have the same pattern, circle *no*.

11. wire, tire, hire yes no
12. flier, drier, crier yes no
13. pure, crier, drier yes no
14. fire, flier, sure yes no
15. sure, cure, pure yes no

Name ____

wire	sure	pure	crier	fire
flier	cure	tire	hire	drier

Match-Ups

Draw a line from each spelling word to its meaning.

1. crier	not dirty or polluted; clean
2. pure	a thin rod of metal
3. wire	a person who cries
4. flier	to make well
5. cure	a person or thing that flies

Sentences to Complete

Write a spelling word on each line to complete the sentence.

6. Will you ______________ me to do the job?
7. The car has a flat ______________ that needs to be changed.
8. Are you ______________ you want to play outside in the rain?
9. My wet towel felt ______________ after it was in the sun.
10. We used logs to make a ______________ at camp.

Name ______________________________

There are six spelling mistakes in the paragraph below. Circle the misspelled words. Write the words correctly on the lines below.

I am flying to visit my grandpa who lives in Italy. The plane should be taking off right now, but it has a flat tiere. I shure hope it gets fixed soon! Now the tire is fixed, but there is a loose wier. This needs to be fixed too. At last, this fliere is ready to take off! I know it can fly in this rainstorm. I think it will be driere when the plane lands in Rome. I hope there is puir sunshine so my grandpa and I can go sightseeing!

1. ____________ **2.** ____________ **3.** ____________

4. ____________ **5.** ____________ **6.** ____________

Writing

Write about visiting a relative who lives far away. Use four words from your spelling list.

____

____

____

____